COMPOUND ENGINES.

TRANSLATED FROM THE FRENCH

OF

A. MALLET.

NEW YORK:
D. VAN NOSTRAND, PUBLISHER,
23 MURRAY AND 27 WARREN STREET.
1874.

COMPOUND ENGINES.

We devote a few pages to the history of Compound Engines; a history heretofore little known, but which we are able to make complete by means of documents that enter into minute details, and which contain matter instructive and interesting, while they shed light upon subjects heretofore obscure.

The idea of employing the expansive power of steam is generally attributed to James Watt. This is shown by the evidence of a patent of Jan. 5, 1769, No. 913. The process consisted in arresting the introduction of steam a little before the termination of the stroke of the piston, thus reducing the pressure at the moment of the reverse stroke; it was not until sometime after that it was perceived that a certain quantity of steam was thus economized.

Jonathan Hornblower, who built the Newcomen engines, patented the use of two cylinders to effect the expansion, on the 13th July, 1781, No. 1298. He said that he employed the steam after its action in the first cylinder in order to employ it in the second expansively.

Here is the original:

"I use two vessels in which the steam is to act, and which in other steam engines are generally called cylinders.

"I employ the steam after it has acted on the first vessel to operate a second time on the other, by permitting it to expand itself, which I do by connecting the vessels together and forming proper channels and apertures, whereby the steam shall occasionally go in and out of the said vessels."

Hornblower's engine met with small success. As it used steam at low pressure it had but a limited expansive power, and the advantages became of no account; rather they became negative on account of the resistances due to the use of two pistons. Besides, he could not use his engine without borrowing most of the parts of Watt's engine, such as the separate condenser, etc. So Hornblower got by means of his invention only the enmity of the friends of Watt, who accused him of indirect plagiarism, and created a bad reputation for him, of which traces are found in the early histories of the steam engine. At this time the use of two cylinders turned out unsuccessful.

But when higher pressure was employed, Woolf did for the engines of Trevithick, Evans, and others, what Hornblower had done for those of Watt; he applied to them the principle of the double cylinder. As he could make use of high pressure, there was promise of success for the invention, and it did succeed, so that he has given his name to engines having two cylinders.

Woolf's patént was taken out in 1804. It contained, as has often been remarked, erroneous notions about the expansive power of steam.

The fact that contributed to the success of Woolf engines, was that although the expansion was not sufficient to yield much advantage over ordinary engines, the division of the work of the steam between the two pistons diminished the differences in pressure and the loss of steam. This was an important matter in the early constructions. Engines of this kind need little repair. We could mention two instances in an industrial centre in Normandy, of engines with two cylinders, which have been in action for nearly fifty years.

In 1805, Willis Earle took out a patent for engines composed of a large and small cylinder superposed, with two pistons mounted on the same rod, a device frequently repeated since that time.

The first Woolf engine was set up in a London brewery. Afterwards Hall made a large number. In 1815 they were introduced into France by Edwards, and they rapidly came into use, without much change in construction. Edwards' engine of 1817 differs hardly at all, even in details, from those that are to-day put up in some of the manufacturing towns. In 1820 the English engineers, Aitken and Steel, built engines with three cylinders, two small and one large. Notice of these engines appears in various works, especially in that of R. Stuart.

In 1824 Joseph Eve patented a compound engine, in which the steam, after acting in a high-pressure engine, passed into a low-pressure engine, where it acted expansively. He employed rotary engines. Here was the first idea of a mode of action different from that of Woolf's engines.

In 1834 Ernest Wolff (a German, we infer, from his name) took out a patent (No. 6,600) of an engine described as compound, as nowadays constructed, which indicates the possibility of modifying existing engines so as to adapt them to the new mode of action.

This patent is very interesting, and it is singular that English authorities hardly refer to it.

It is certain that compound engines with two cylinders and intermediate reservoir, to which the name of Woolf has been given, though they have not the same mode of action, should be called "Wolff engines."

We give the essential part of this patent. "The invention consists of the combination of two or more engines, each complete in all its parts, and so disposed that while the first receives steam at one, two, or more atmospheres of pressure, the next engine is moved by the steam that escapes from the first. In the last engine the steam is condensed in the ordinary way, or escapes into the atmosphere. The work supplied by the several engines is applied to the same shaft, or to several combined, or to independent shafts.

"As in steam vessels and other applications, two conjoined engines are generally employed. The present invention is especially adapted for this purpose, as it presents economic advantages; as it reduces the expense of the apparatus without increasing its complication.

"It is sometimes useful to have between the cylinders an intermediate reservoir to regulate the pressure; this may be placed with advantage at the base of the chimney, so as to maintain or raise the temperature and the pressure of the steam in its passage from one cylinder to the other. Indeed, if necessary, the heat may be supplied by a special fire-box.

"It is often necessary to employ a special pipe with a stopcock to admit the steam from the boiler to an intermediate reservoir in order to give to the machine the power of starting any crank. This direct introduction may be employed to increase for a time the power of the engine."

The writer then explains a method of modifying old engines by adding to a high-pressure engine a low-pressure cylinder; or, in the case of a marine engine, by substituting for one of the low-pressure cylinders a high-pressure cylinder.

The drawing annexed to the patent shows a pair of marine beam-engines.

In 1837, William Gilman patented an engine consisting of two cylinders placed one on the other, one of them having an annular piston with a single cut-off, with multiple ports disconnecting the two cylinders. This disposition has been often reproduced, and is frequently employed nowadays, especially in Sweden. Gilman also describes an engine of three cylinders in which the steam acts in succession.

In 1837, Jonathan Dickson patented (No. 7,439) engines in which the steam acts successively by means of boilers with decreasing pressure, or parts of boilers constituting a compound boiler. This contrivance has also been made use of since the time of the invention. In fact, it is nothing more than Woolf's patent: for this proposes to re-heat the intermediate reservoir by a special fire-box, a process which constitutes in a certain way a low-pressure boiler.

Dickson proposes the use of feed-pumps to serve as guides to the piston cranks, and to control the slide-valves of each engine by the other engine.

In the same year James Slater patented (No. 7,467) engines acting in the same way, with an intermediate reservoir, employing a low-pressure boiler. He describes a regulating valve designed to keep the steam pressure at a fixed point, and also to start the engines. This is nothing more than Woolf's invention—the valve, perhaps, excepted.

The drawings annexed to the patent show various applications, especially a pair of marine beam-engines.

William Whitman, in 1839, patented an engine in which the piston has a sheath on one side only, so that the cylinder has two different capacities. The steam first acts in the annular space, then expands into the other portion of the cylinder. This disposition, applied with some success by the inventor, has been frequently reproduced. It is probably the simplest way of applying the Woolf method of action.

In 1841, James Sims patented an engine of two superposed cylinders, with pistons on the same rod; with this special distinction, that the bottom of the smaller piston is in constant communication with the top of the larger.

In 1842, Hinrik Zander took out a patent (No. 9,516) of an engine, in which the steam acts in the first cylinder expansively, to a certain extent, then passes into two others which are larger, and expands. The three cylinders are attached to the same shaft, so that their motion may be as uniform as possible. The low-pressure cylinders are provided with jackets which contain the steam from the boiler. Zander describes intermediate reservoirs, and proposes to introduce into them, or into a communicating pipe supplying their place, a float-valve to allow the escape of condensed water.

In his drawings is represented a disposition in which the crank of an oscillating high-pressure cylinder, placed obliquely, is attached to the crank of a marine beam engine. This engineer (probably of Holland) seems, according to documents which we have found, to have built some marine engines on this plan.

Octavius Henry Smith patented, in 1844, an engine acting on the Woolfian principle, consisting of a high-pressure and a low-pressure cylinder, both oscillating and having their rods attached to the same crank. The Cricket Engine, referred to in Bramwell's memoir, is of this kind. A complete description will be found in the "Practical Mechanic and Engineer's Magazine," 1847. Afterwards, we find many patents of expansion engines. We mention only those of Perkins, 1844; McNaught, 1845, which modified old engines by the addition of a high-pressure cylinder; of Thomas Craddock, 1852; Daniel Adamson and Leonard Cooper, 1852, which superheated the steam in its passage from the high to the low-pressure cylinder, by means of tubes set in the smoke-box of a tubular boiler.

We shall not go further in our examination of these patents. It is perceived that, since 1852, all the essential elements of the action of steam by expansion, in separate cylinders, have been pointed out, and that there remains nothing to be invented, even in perfecting details. We shall look further back for applications.

The Cricket Engine was built in 1847, by Joyce & Co., of Greenwich. It exploded the same year. Bramwell speaks of a boat built at about the same time by Spiller, in which was placed an engine, consisting of a low and high-pressure cylinder. We have found no document concerning it.

According to the authority of "Zeitschrift des Oster Ing. and Arch.," 1867, M. Roetgen of Rotterdam has built engines, since 1840, composed of cylinders inclined towards each other, and acting on the same pair of cranks, the same steam being successively used in the two cylinders. These engines were put into the boats Elizabeth, Stadt Magdeburgh, Kron-Prinz Paul Friedrich.

We do not regard the date 1840 as exact. If, as is probably the case, these engines are those made according to the plans of Zander, they were evidently built after his patent of 1842.

The journal "Engineering" of Sept. 9, 1870, contains a description and drawing of an engine built in 1848 by the Sterkerader Hütte for the Rhine boat Kron-Prinz von Preussen. This engine had two cylinders one 0^{m},508 in diameter, and 0^{m},800 long; the other 0^{m},914 in diameter, and 0 ,914 in length. Each acted on a crank; the two cranks were connected so that the effect was the same as if the cylinders acted at right angles upon the cranks, while the angle between the axes was 130°. There was no special intermediate reservoir. The connecting-pipe 0^{m},254 in diameter, acted in its stead. There were no steam jackets, and, as no precaution was taken to prevent the condensation of steam in its passage from one cylinder to the other, economical results could not be expected.

Still it is a fact that Feyenoord's works at Rotterdam, where these engines were first constructed, have never given them up. We ourselves saw at Rotterdam in 1860 a steamboat of 70 h. p. nominal, the Wilhelm II., which had served as a pleasure boat for the King of Holland. The engine, with low-pressure cylinder, had been modified by the addition of a high-pressure cylinder inclined to the other, acting on the same crank, the same steam working successively in the cylinders.

It would be unjust to omit mention of Carillon, a Paris builder, who succeeded (1842) in making a low-pressure engine work with the discharged steam of one at high pressure. This was set up at the St. Louis Glass Works. The essay seems not to have been repeated; being abandoned, we think, because of the failure of a surface condenser.

In 1852, James Samuel applied the principle of continuous expansion to locomotive engines. This consists of a simultaneous action of steam upon the two pistons. Suppose two pistons whose rods act at right angles to the crank. The steam works at full pressure during half the stroke of the first. At this moment admission ceases, and the first cylinder is put into communication with the second while its piston is at the beginning of its stroke. Expansion occurs simultaneously in the two cylinders until near the end of the stroke of the first, then in the second, only till nearly the end of its stroke.

This system, related to that of Milner, mentioned in Bramwell's memoir, has been again taken up by Stewart & Nicholson, and applied in the tugs on the Thames. Though simple, it has the disadvantage of not avoiding great depression of temperature as well as those of Woolf & Wolff, since the two cylinders communicate with the discharge ports or the condenser.

The experiments of Samuel, reported in the "Memoirs of the Institution of Mechanical Engineers," 1852, were made on a freight and a passenger engine on the Eastern Counties Railway. In the first there were two equal cylinders; in the second the larger cylinder had a section twice as large as that of the smaller. Though the results seemed quite favorable, the essays were abandoned until the time when they were again resumed by Stewart & Nicholson. This is inferior to the other kinds of compound engine.

The first noted applications of double-cylinder engines were made at Glasgow, in 1856, by Randolph & Elder. A little after, Rowan & Horton constructed three cylinder engines; one high-pressure feeding two others. There were 6 cylinders in the machine. The steam was supplied at a pressure of 8 atmospheres by boilers of a special form.

One of these engines, according to Rankine, should not consume more than 0,500 k. of fuel per horse power hourly. This would seem doubtful; but it would be useless to discuss the point, for the engines have not stood the test of service. The boilers are rapidly destroyed, and the construction is too complicated. The condensers were surface condensers of a particular pattern.

Rowan & Horton put one of their engines into L'Actif, a French vessel.

In 1859, Humphreys & Tennant, of Deptford, built for the Peninsular and Oriental Company Woolf engines with moderate tensions. These engines, set up in the steamers Poonah, Mooltan, Carnatic, Baroda, Delhi, etc., at first gave good results. The pressure was 25 lbs., with surface condensation. The cylinders of the Mooltan were 96 and 46 in. in diameter, with a length of 3 ft. Consumption was 2½ lbs. per horse power. The good results were not permanent, especially in matters of detail.

All these have been replaced by single engines built by Humphreys & Tennant.

In 1861, Normand changed to the Wolff the engine of the small steamer Le Furet built by Penn. The engine worked at 6 atmospheres with intermediate reservoir, re-heating, and *monhydric* condensation. The results were excellent. Afterwards Normand altered in the same way the engines of the Eclair, the Albert, etc., and still constructs the same kind of engines.

The Imperial Marine made essays moderately successful with three cylinders. The expansion was not great enough, the cylinders being of the same diameter, so that the economic advantage was not important.

But the principle seems natural, and the English Admiralty is at present changing the engines of the ship Jumna.

Escher, Wyss & Co., of Zurich, have built, from the plans of their engineer, Murray Jackson, marine engines with a low and a high-pressure cylinder, set side by side and acting perpendicularly to the cranks. One of these engines was exhibited at Paris in 1861, but it was out of sight under a shed. They have no special intermediate reservoir, the connecting pipe of the cylinders acting in its place. This firm have constructed a large number for the Swiss and Italian lakes, for the Danube, Rhine, and other rivers. Their engines are of the Woolf system. One with four cylinders was exhibited at London in 1862; it is now upon a boat upon Lake Lucerne.

The compound marine Wolff engine is at present built in many English shops;

though some maintain the Woolf type, with superposed cylinders. In France all engines are of the first kind.

II. We now consider the method of action in compound engines, beginning with those of Woolf.

In this system the pistons almost always move parallel and in the same direction, although engines with opposing cranks have been constructed (Randolph & Elder; Bondier Frères, of Rouen; Carret, Marshall & Co.) in order to have more direct connection between the cylinders. We suppose that the cylinders have the same length.

The steam acts directly upon the first piston, then expansively; and when the small piston is at the end of its stroke, the connection with the large begins, so that the space under the action of steam at any instant is composed of a fraction of each cylinder, the fractions having an inverse ratio.

We calculate the volume for each period of the stroke, and find the corresponding pressure by Mariotte's law. Let S and s represent the areas of the pistons, l the stroke when the pistons are at a distance z

from the beginning of the stroke. The remaining volume in the small cylinder is $s(l\text{-}z)$, the volume of steam in the large cylinder is Sz, hence the total volume occupied by the steam is

$$s\,(l-z)+S\,z$$

or

$$s\,l-s\,z+S\,z=s\,l+(S-s)\,z.$$

If P′ is the pressure at the end of the stroke of the small piston, the pressure at any point between the two pistons is

$$X=\frac{P'\,s\,l}{s\,l+\,(S-s)\,z}.$$

By giving to z a number of values, the curve of expansion may be constructed.

P′ is the boiler pressure if there is no expansion in the small cylinder; otherwise, from P′ we deduct P′ from the pressure P in the boiler by the relation

$$P'=P\,.\frac{l}{l\,n}.$$

n being the expansion in the first cylinder. Denoting by v and V the volumes of the cylinders, by m the total expansion, we have for the expenditure of steam

$$q=\frac{V}{n}.$$

The work of the volume q of steam is

$$T = qP\left(1 + 2,3026 \log. \frac{V}{q}\right)$$

an expression which does not contain v, *i. e.*, the work does not depend on the volume of the small cylinder, but only on the volume q of steam expended, and upon the dimensions of the large cylinder and the initial pressure.

Theoretically, then, the work is the same as if there were no small cylinder, and the volume of steam introduced into it is directly expended in the large cylinder. Should we therefore conclude, as most persons do, that the small piston does no work, and that it is merely a distributor? This would be utterly erroneous; the work of the low-pressure cylinder alone, would be represented by V and by a certain mean pressure π. Then denoting by p the mean pressure in the small cylinder, by p' the mean pressure between the two pistons, we have

$$T = v\,p - v\,p' + V\,p.$$
$$= p\,v + p'\,(V - v) = V\,\pi.$$
$$\therefore \pi = p' + \frac{v}{V}\,(p - p'),$$

a value always larger than p'; hence the total work is always greater than that of the large cylinder.

From the expression $m = \frac{V}{v} n$ we have $\frac{V}{v} = \frac{m}{n}$, which determines the proportion between the cylinders for a given total expansion and introduction into the first cylinder.

We observe that in order to have $V = v$, or equal cylinders, we must have $n = m$; and the second cylinder would be useless, so that the two cylinders could be practically equal in Woolf's engine. The same is true in the Wolff engine, as will appear.

The following table shows the dimensions of the large cylinder (those of the small being equal to unity) for given total expansions and admissions into the small cylinder:

Admissions to the small cylinder.		0.3.	0.4.	0.5.
Total expansion	5	1.5	2	2.5
	10	3	4	5
	15	4.5	6	7.5
	20	6	8	10.

Admissions to the small cylinder.		0.6.	0.7.	0.8.	0.9.
Total expansion	5	3	3.5	4	4.5
	10	6	7.	8	9
	15	9	10.5	12	13.5
	20	10	14.	16	18

The ratio 5 is not exceeded in practice; it being better, for large expansions, to increase the expansion for the first cylinder.

We have so far supposed that there was no dead space, a condition never realized. This space is composed of two parts; one, the space between the small piston and the bottom of the cylinder at the end of its stroke and the port of the small cylinder, the other composed of the interior capacity of the small slide valve, of the connecting pipe between the valve boxes (we suppose each cylinder has its special distributor), of the valve box of the large cylinder, of the port of the large cylinder, and finally of the free space between the large piston and the bottom of the cylinder at the end of its stroke.

The first space contains, at the end of the stroke, an amount of steam (P′) of the final tension of the stroke of the small piston; but the second space contains steam of a tension (P″) corresponding to that existing at the end of the stroke of the large piston; and it should be filled with steam of the tension P′.

A communication is established between the two cylinders; the tension diminishes because of the dead space not corresponding to the displacement of the pistons; and the mean pressure on the large piston is considerably diminished.

In the old two-cylinder engines the dead space is considerable, sometimes exceeding one-third the volume of the small cylinder; but it has been diminished by a suitable disposition of slide-valves; and it can be prevented in a certain measure by causing *compression* at the end of the stroke of the large piston.

If the dead space is a fraction $\frac{1}{k}$ of the volume v of the small cylinder, the pressure at the beginning of the stroke of the large piston, instead of being $\frac{1}{k}$ will be $\frac{1}{V + \frac{v}{k}}$. For example, for $k = 3$, it would be

$$\frac{1}{1.33} = 0.75,$$

instead of 1.

If the initial pressure in the large cylinder, instead of P′ is $\frac{P' v}{v + \frac{v}{k}}$, at the end of its stroke, when the volume is $V + \frac{v}{k}$, the final tension will be

$$\frac{P' v}{v + \frac{v}{k}} + \frac{v + \frac{v}{k}}{V + \frac{v}{k}} = \frac{P' v}{V + \frac{v}{k}}$$

instead of $\frac{P' v}{V}$. The ratio of the two volumes is $\dfrac{1}{1+\dfrac{v}{V k}}$. If $\dfrac{v}{V}=0.2$ and $k=3$, the ratio is 0.98.

The final tension is not sensibly modified because the ratio of the volume of the cylinders diminishes the effect of the dead space; that is, because the steam contained in this dead space expands during the stroke and tends to restore the tension.

For a better understanding of the subject, and to show the action of the engines, we reproduce in Figs. 1 and 2 the diagrams of indicator of a large Woolf beam engine, built by Slawecki, an engineer at Rouen, who subjected the engine to interesting dynamometric experiments.

Fig. 1 shows the curves of the two cylinders to the same scale and with the same atmospheric line. We observe that the counter-pressure of the small cylinder corresponds very exactly to the pressure upon the large piston, showing that the resistances in the passages between the two cylinders are so diminished as to be almost insensible.

That the two curves coincide at one point, is due to the fact that they were taken at

FIG. 1.

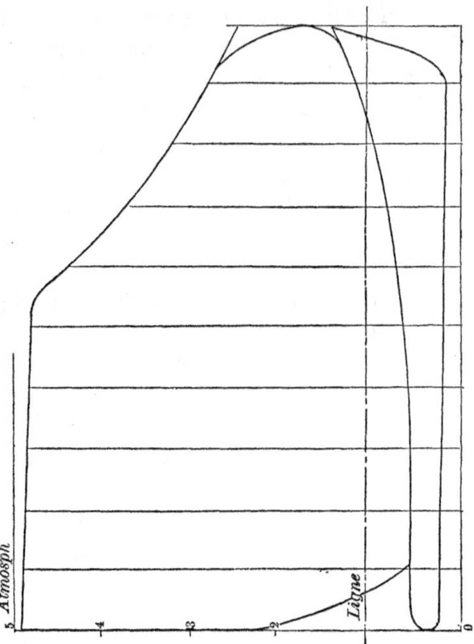

the top of the cylinder, where the real representation of the action of the *same* steam upon each piston is not given.

The effect of dead space is shown by the difference between the final pressure in the small cylinder and the initial pressure in the large. In this case the difference is one and a half atmospheres.

In order to compare the action of the Woolf with that of an ordinary engine with a single cylinder, we show in Fig. 2 two curves to the same vertical scale for ordinary pressures, but with abscissas to scales proportional to the volumes of the cylinders. As the two areas coincide upon the portion marked with vertical hatchings, this portion has been transferred to an equal area hatched horizontally. The curve of work of the steam acting on a single cylinder with the same expansion as in the engine with two cylinders is given; and the small portion of the area corresponding to the work of the small cylinder beyond curve is shown within it.

It is seen that the final pressure is exactly the same in the two systems, except that there is a defect of area of work in

FIG. 2.

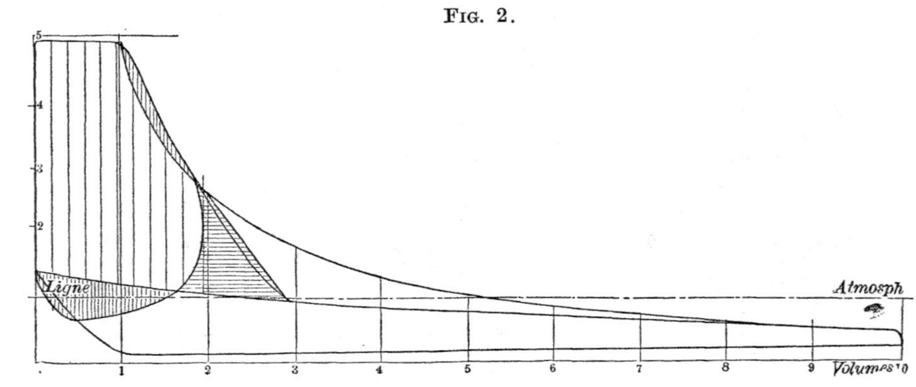

Woolf's engine corresponding to the loss due to dead space. Calculating the area of this portion, we find the loss due to the use of two cylinders to be nearly 15 per cent.

The majority of authorities rest here, and take this result as a text for condemning the compound-engine, at least in principle. We shall see that the *physical* loss in the action of expansive engines with one cylinder far surpasses that of the Woolf, a loss in some sort apparent, and depending upon a simply geometric cause.

This would be the place to examine the causes of the superiority of this engine due to expansion; but as this superiority is common, in theory at least, to all double-cylinder engines, we shall consider the question after an examination of the second kind of engines.

We close with this remark concerning Woolf engines: if each cylinder is provided with a special distribution, it is not necessary that there should be a mathematical correspondence of exhaust port of the first cylinder with the admission port of the second. The existence of dead spaces permits separation by a certain interval; while near the dead-points the angular displacement of the crank corresponds to a very small linear displacement of the piston. Hence if each cylinder acts upon a special crank, the two cranks may be set at an angle of 45 deg. to 135 deg., acccording as the pistons are to move in the same or opposite directions. This disposition which facilitates the passage of the dead-points is sometimes employed, as appears in Bramwell's memoir.

III. In the second system of engines, which should be called Wolff engines, the second cylinder is not supplied directly by the first, but by an intermediate reservoir of such dimensions that the pressure within it may be regarded as nearly constant, being in some sort an engine with graduated pressure.

Denoting by P the pressure in the boiler, supposed equal to pressure upon entering the first cylinder, and by P′ the pressure in the intermediate reservoir, due in the first place to direct introduction of the steam from the boiler, and controlled by a safety valve; by P″ the resisting pressure at the discharge port or at the condenser; then, if the expansion n in the first cylinder is such that the final pressure in this cylinder is P′, or differs from it but little (v being the volume of the small cylinder, and q the volume of the steam admitted), we have

$$\frac{P'}{P} = \frac{q}{v} = \frac{1}{n}.$$

The second cylinder receives all the steam from the boiler at the pressure P″.

Then $\frac{V}{q} = \frac{P}{P'} = m$; m being the total expansion, hence $\frac{V}{v} = \frac{m}{n}$, and the ratio of the volumes of the cylinders is the same as in the first system. But the large cylinder, instead of receiving the steam during the entire stroke, receives it only while a fraction equal to the expansion n' corresponds to the decrease of the pressure from P′ to P″; $\frac{P'}{P''} = n$ and since $\frac{P}{P'} = n$ and $\frac{P'}{P''} = n'$, $\frac{P}{P''} = n\,n' = m.$

The total expansion is equal to the product of the partial expansions, and the expansion in the second cylinder is equal to the ratio of the volumes of the two cylinders. We see that if one should make the two cylinders equal in an engine of this kind, there would be no expansion in the second cylinder. The engine might act if there were a notable difference between P′ and P″, but the whole possible expansion would not be utilized, and the system would have no *raison d'être.*

One of the advantages of this arrangement is that the dead spaces are not hurtful, as they are actually utilized by enlarging them.

The work of the steam in the first cylinder is

$$T = q \cdot P\left(1 + e \log \frac{V}{q}\right) - P' v.$$

that in the second is

$$T = q' P' \left(1 + \log \frac{V}{q'}\right) - P'' V.$$

Taking the sum for the total work, and substituting

$$q = \frac{V}{m},\ q' = V\frac{n}{m},\ p' = \frac{P}{n},$$

we have

$$T = \frac{V}{m}\left(P\,(1 + e \log n - 1 + 1 + e \log \frac{m}{n}\right) - P''m),$$

$$T = \frac{V}{m} P\,(1 + e \log m) - P''\,m$$

$$= \frac{V}{m} P\,(1 + e \log m) - V\,P''$$

$$= q\,P\left(1 + e \log \frac{V}{q}\right) - V\,P''.$$

This expression represents the work in the large cylinder of the quantity of steam at tension P, expanded from volume q to the volume V.

The principle stated concerning Woolf's engine also holds as to the action of this kind; that is, the work is the same as if the large cylinder were the only one, and the steam admitted to the small cylinder were directly introduced into the large, and expanded in the ratio of the total expansion.

In this kind of engine the work of the two cylinders is regulated as much as possible. Equating the expressions of work given above, we find after proper substitutions

$$v\left(\frac{P(1+e\log n)}{n}-P_{/}\right)=V\left(\frac{n}{m}P'(1+e\log\frac{m}{n})-P''\right),$$

and substituting for P′ its value $\frac{P}{n}$,

$$\frac{V}{v}=\frac{\frac{P(1+e\log n)-P}{n}}{\frac{P}{m}\left(1+e\log\frac{m}{n}\right)-P''}=$$

$$\frac{\frac{P}{n}e\log n}{\frac{P}{m}(1+e\log n')-P''}$$

This expression gives the ratio $\frac{V}{v}$ for known expansions n and n', for we always have $m = n\,n'$.

Remembering that

$$\frac{P}{n}=P' \text{ and } \frac{P}{m}=P''$$

we have

$$\frac{V}{v}=\frac{P'\,e\log n}{P''\,e\log n'}=n'\,\frac{\log n}{n'}$$

and for $n = n'$

$$\frac{V}{v}=n$$

a relation already known.

In Woolf's engine the extreme difference

of temperatures in the small cylinder is larger than in the engine with intermediate reservoir. Here is a theoretic inferiority, at least for engines of slow action in which the decrease of temperature due to the diminution of tension is noteworthy. Hence, as we shall see, the use of an envelope of steam is in this case more necessary.

If the engine with reservoir has a real advantage over that of Woolf, it has, on the other hand, the inconvenience of requiring for considerable expansions, that the admission be early cut off in each cylinder: since for a total expansion of 10 volumes the expansion should occupy $\frac{1}{5}$ in one cylinder, $\frac{1}{2}$ in the other, or $\frac{1}{3.16}$ in each, if expansions in both are equal; and we know that ordinary distribution does not favor small introductions.

Generally a fixed expansion is imposed upon the large cylinder; and the small cylinder is provided with a special apparatus, with variable cut-off to reduce admission as desired. It is difficult, however, to effect large expansions, and too great expansion in the large cylinder produces some of the disadvantages of the single cylinder. It would be better, in this case, to employ multiple expansions and several successive cylinders, as has often been proposed.

We shall now investigate the causes of the physical superiority in the action of compound engines over those with a single cylinder.

These causes are two in number: The first, which is common to all engines of this class, is due to the fact that the difference of extreme temperatures in each cylinder is less, and that the interior condensation is much diminished. The second, which is peculiar to engines with an intermediate reservoir, is due to the partial removal of the water from the steam which held it in suspension at the time of its leaving the first cylinder; so that the water does not pass, in a liquid form at least, to the second cylinder.

For the better comprehension of these effects, it is necessary first to examine what takes place in single-cylinder engines. Suppose an engine of this kind acting expansively; the steam is admitted during, say one-fifth of the stroke. Then the piston moves by virtue of the expansion, and the pressure diminishes in a certain ratio with the increase of volume, until the end of the stroke. As the temperature of the steam diminishes with its pressure, a contraction takes place in consequence of the cooling, which acts at the same time with the expansion to diminish the pressure; and we perceive that if the steam receives no heat during its expansion, the pressures will diminish more rapidly than by Mariotte's law. But, on investigation of the action of expanding engines without cylinder envelope, we find a different condition of things. We have taken a number of indicator diagrams from a non-condensing engine, into which admission took place only during very small fractions of the stroke, varying from $\frac{1}{40}$ to $\frac{1}{8}$. Besides, the actual expansions were much less because of dead space; the correction reducing the apparent expansions from 40, 20, 13.33, 10 and 8 to 14, 10.5, 8.04, 7, 6 volumes.

The figure (3) shows the diagram from

an engine with actual expansion of 14 volumes. The initial absolute pressure being 2.75 atmospheres, the tension at the end of the stroke, by Mariotte's law, should be considerably below the atmospheric pressure. The diagram shows the contrary. The pressure, which decreases rapidly near the post, afterwards approaches the horizontal line, and at the end of the stroke is considerably above the atmospheric line. This is due to the vaporization of the water in the cylinder during the period of expansion, which furnishes a supplement of steam whose tension is added to the primitive tension.

At first it seems that it is an advantage that the work obtained is greater than that due to Mariotte's law; but we must count the cost, and it is easy to show that it costs as much or more than the work of full pressure during the entire stroke, so that all advantage of the expansion is lost.

By calculation of the area of the curve and of the mean ordinate, we shall find the mean effective tension of the steam to be 0.236 kil. per square centimetre. The cyl-

Figs. 3 and 4.

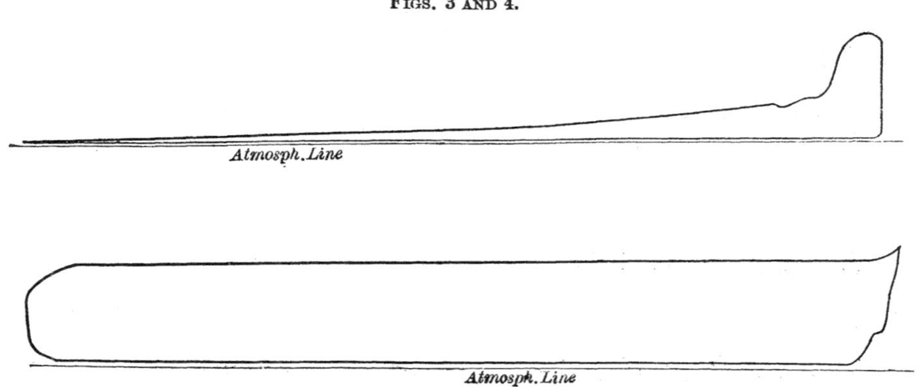

inder has a diameter of 0.200 m., a stroke of 0.40 m., 30 strokes per minute; the work on the piston is 29.2 kilogrammetres. The volume of steam in the cylinder at the instant of closing of the admission port, dead space included, is 0.942*l*. This weighs (at a tension of 2.75 at.) 1.50 gr. per litre; that is, 1.163 gr. for 30 strokes per minute, corresponding to 4,547 per hour, or 11.65 H. P. of sensible steam acting on the piston.

If we now take a volume corresponding to $\frac{1}{10}$ of the stroke, and, assuming that the steam is in a saturated condition in the cylinder, apply to this volume the weight per litre corresponding to the observed pressure, we find at this moment in the cylinder a weight not of 1,263 gr., but of 1,410 gr.; at the middle of the course it is 4,837 gr. At the end it is still less, but at this point observation becomes more difficult because the pressure comes near the atmospheric line. We assume the figures 4,837 gr., which correspond to a sensible expenditure of 47 kilog. of steam per H. P., hourly; *i. e.*, 4 times the amount found above.

Referring to figure answering to the admission for the whole stroke, we find that the mean ordinate of mean effective pressure is 1.56 kil., giving for 30 revolutions a work on the piston of 196 kilogrammetres, or 2,615 H. P. The volume of steam which fills the cylinder at the end of the stroke is 12.56*l.*; which, at a pressure of 2.95 atmospheres, gives 18.84 gr., or 67.824 kil. hourly, or 26 kil. per H. P.

As there can be no other steam in the cylinder, we must conclude that the engine uses less at full pressure than when acting with full expansion, a fact long ago verified.

The presence of the great quantity of steam in the cylinder, can be explained only as due to vaporization during the period of expansion of the water contained in the cylinder, which is caused by diminution of pressure, and at the expense of the heat of the metal. But from what source does this water come? In the case under consideration the water was supplied by a tubular boiler; but the engine went very slow and did very little work.

As the dimensions of the boiler corresponded to a performance 15 or 20 times as great, it is difficult to admit that any considerable portion of the water was held in suspension in the steam. It was almost entirely due to the condensation at the admission port. The steam meets the surfaces, walls and ends of the cylinder, the piston and the rod, all of which are at a lower temperature, and partly condenses, while it raises the temperature of the metal. To raise 10 kilog. of metal 10 deg. requires $10 \times 10 \times 0.15$, or 15 units of heat, answering to a condensation of about 30 grammes of steam. Hence, at the beginning of the stroke there is a certain quantity of water; it is this which vaporizes as soon as the pressure at the admission port diminishes, by absorbing the heat of the walls of the cylinder and that of the piston, a heat which is lost by escape or by transfer to the condenser. The cooled metal then demands from the steam just arrived from the boiler a fresh quantity of heat, acting as an agent to exchange between boiler and condenser. This explains the fact often noticed, that in condensing engines, having cylinders without jackets, it is more difficult to effect a vacuum with complete expansion than with full steam.

It is to be observed that the elevation of the temperature of the walls of the cylinder by condensation does not take place instantaneously, any more than does the vaporization that attends expansion. Time has its part in the phenomenon. In engines of slow action the loss of heat is much greater than in those of more rapid movement. The engine in our own experiments was under the most unfavorable conditions; acting without condensation, at full expansion, slowly, and at a low pressure.

The office of steam jackets is to keep the walls of the cylinder at a constant temperature, so as to prevent the presence of water in the cylinder and the resulting inconveniences. A certain quantity of heat is lost, corresponding to the condensation in the jacket; but as the pressure in the jacket does not vary, this condensed steam will remain in the condition of water, or will be expelled by the clearing valves; it will not pass back into the state of vapor while absorbing the heat, as would be the case if it had been condensed in the cylinder. This is the true statement of the action of jackets: *they maintain the heat in the cylinder and cause condensation to take place in the jacket where the pressure is constant, and not in the cylinder where it is variable.*

The jacket does not always prevent the formation of water within the body of steam, because of very prolonged expansions due to cooling and dilatation. We shall find it possible to avoid these consequences.

The following observations are of interest:

(1) We note the little efficacy in the action of jackets of wood, felt, polished brass, etc., which resist only cooling from without.

(2) It is easy to show that little would be effected by replacing the steam in the jacket by the heated gases of combustion, as has been attempted in some engines (sometimes the interior cylinders of locomotive engines are placed in the smoke chest).

A cubic metre of steam at 5 atmospheres weighs 2,600 kil., and by condensation may lose 500 heat units per kilogramme, or 1,300 per cubic metre. A cubic metre of gas from combustion, at a temperature of 350 deg. represents 0.66 kil., giving $200 \times 0.25 \times 0.66 = 33$ heat units.

To replace a layer of steam a centimetre in thickness would require a layer 40 centimetres thick, of heated gas (assuming that the conductibility is perfect), or 40 square metres to replace a square metre of heated steam.

(3) In the old engines the steam from the boiler circulated in the jacket before entering the cylinder. It is easy to see that this would cause steam to pass into the cylinders containing water in suspension, which would cause a clear loss of heat in case of expansion; then this water would remain in the jacket without disadvantage. A special pipe should always be used through which to pass the steam into the jackets.

What has been said regarding the jackets applies equally to the cylinder ends and to the pistons. The importance of regarding the latter is great in engines of short stroke, in which the diameter is 2½ times or more than the stroke. In these engines the ends are always heated; and pistons are heated by means of grooved rods and other devices.

Nowadays cylinders and their jackets are generally of the same coating, in order to avoid complex and delicate adjustments. Hence serious difficulties in construction. Perkins, in the engines of the steamer Filga, substituted for the jacket a serpentine iron tube sunk into the cylinder. In this tube the steam circulates. This compels an increased thickness of the cylinder. But the weight need not be much greater than with a jacket, and the construction is easier.

We return to two-cylinder engines. Notwithstanding the real efficacy of steam jackets, still it is certain that there is always more or less condensation of water during expansion. This is necessarily proportional to the extreme differences of temperature to which the cylinder is subjected, as well as to the extent of cooling surfaces.

In double-cylinder engines, especially with intermediate reservoir, this difference of temperature is reduced for each cylinder. If we suppose a pressure of 5 atmospheres, corresponding to 152 deg., an intermediate pressure of 2, corresponding to 120 deg., and a final pressure of 0.50, corresponding to 81 deg., the difference in the first cylinder will be 32 deg., in the second, 39 deg. With a single cylinder the difference would be 71 deg.

We shall find that the total condensation will be considerably diminished. Suppose the ratio of volumes of the two cylinders to be 2, 5, the first being 0.78, its total cooling surface is 4.70. The volume of the large cylinder being $0.78 \times 2.5 = 1.95$, its cooling surface is 8.86. We have then

$$4.7 \times 32 + 8.86 \times 39 = 495.$$

A single-cylinder engine of the same work and the same expansion should have the dimensions of the large cylinder. In this case we have

$$8.86 \times 71 = 629.$$

The advantage in favor of the double-cylinder engine is 21 per cent. It is so real that, as mentioned before, jackets are sometimes dispensed with in compound engines, when the expansions are not great. It is possible also in engines with intermediate reservoir, which have the special advantage, that the water held in suspension at the time of leaving the first cylinder is deposited in the reservoir and does not pass into the second cylinder. It would seem that it always has been observed that the intermediate reservoir produces much water, for Zander, in his patent, mentions the use of a float valve to discharge this water. The intermediate heater vaporizes this water and makes it do work again in the large cylinder.

The following are the conclusions which we think can be drawn from our investigations:

(1) The usefulness of a steam envelope is incontestable, being greater as the differences in temperature increase, so that jackets are more advantageous for condensing than for non-condensing engines, for great

expansion than for slight, for single than for double-cylinder engines. But when expansion is considerable, the jacket is not sufficient.

(2) Engines of two cylinders have a decided superiority over those with one, so that for moderate expansions, steam jackets may be dispensed with.

(3) The compound engine, which we have called the Wolff, has advantages over the ordinary Woolf engine, because the dead spaces have less influence, and because the steam that has worked in the first cylinder can be there separated from the water, so as to work better in the second cylinder. Again, the small cylinder is less exposed to interior cooling, and it is less necessary to employ the steam jacket.

These facts seem very simple, yet they are often misconceived. For example, we read in the work of a distinguished author, as follows: "The means by which economy of steam is attained, consist, while using mean or high pressures, (1) in superheating the steam; (2) in employing long expansions, either by Woolf's method, or directly.

This requires, all other things being equal, less complicated apparatus, and it is considered efficient at least as regards the utilization of the steam. In land engines, where space is of little account, some builders propose to make use of expansions of 0.95 (20 volumes). In all such cases the cylinders have envelopes. But the use of jackets with circulation of steam is to be avoided."

This was written in 1866. We think that, in view of a great number of facts and of the general use of the compound engine, the author should modify his conclusions.

The Bulletin of the Industrial Scientific Society of Marseilles (1873) contains this note upon steam jackets, by Stapfer:

"Steam jackets are a costly addition, which only apparently increase the power of an engine. They seldom last more than two years, unless with the greatest precaution. Indeed, there are few engines in which they are not much obstructed. Of course, I do not refer to locomotives whose cylinders are inside the steam chest or in the smoke box, but only to return-water jackets. It is obvious that by placing the cylinder in an atmosphere of waste gases at 300 or 400 deg., a very appreciable quantity of heat can be utilized. But these are conditions seldom realized, and they belong rather to the construction of the boiler."

We can here only refer the reader to what has been said above concerning the real value of the direct action of the gases of combustion. They can be turned to account only by the aid of considerable surfaces. Those of the cylinders and the intermediate reservoir are not sufficient.

The heat of the gases of combustion has often been employed to reheat the steam in its passage from one cylinder to the other. Normand has made use of a tubular apparatus set in the smoke box. This was applied to his first engines and is found in his latest, as in the Ville de Brest and the Belgrano.

Stapfer says that jackets use too much steam; and to a certain extent without useful result, by superheating the steam at the end of its course in the large cylinder, when it is about to pass to the condenser. This objection has probably in some cases caused the suppression of the jacket of the large cylinder. Though the action of the jacket may be very efficient in vaporizing the water of condensation, especially on the cylinder walls, it is of little effect in heating the steam, the conductibility of which is very feeble. If, then, there is a superheating of steam, it occurs only in immediate proximity with the walls, and only when all the water has been vaporized. This vaporization and the effect of the jacket are of no account while the piston is driven by the steam. That the loss should be appreciable particular conditions are necessary; for example, that the escape should take place during a considerable portion of the stroke.

In engines of rapid action, the transmission of heat would not be sufficiently rapid. Stapfer seems to perceive this fact, for he says: "Steam jackets would be good for engines of slow action, in which heating takes place slowly."

The objection regarding the waste of steam by the jacket, made even when it is proven that this helps to avoid a greater waste, holds true in a degree only. But it is worth while to reduce it as much as possible. We think this can be done by employing the gases of combustion, not in directly heating the cylinders, but in generating steam to feed the jackets. The apparatus might consist of a small group of tubes at the base of the chimney where a portion of the water condensed in the jackets would collect, either by the action of its own weight or by contrivances easily adjusted.

The water would always be the same, and not coming into contact with impure substances, it would remain free from the obstructions.

The steam generated in this small boiler heated by the gases of combustion would serve only to heat the cylinders, and not at all in the direct production of motive power; so that it would be possible to supply the heat necessary for the jackets without cost.

Attempts have been made to utilize the heat held in the steam at its discharge from the cylinder by vaporizing a liquid more volatile than water, and to have this vapor act upon a second piston. By this means more work would be realized without employing much of the expansion of the first cylinder. These engines, invented by Dutrembley, who has made a great number of them, are engines of graduated pressure (*pressions étagées*). It is easy to show that the same result is obtained more simply by utilizing the expansion of the steam in successive cylinders.

Suppose a kilogramme of steam leaves the first cylinder and enters the ether condenser at 80 deg. or at one-half an atmosphere. This steam can supply 550 heat units and will vaporize about 3.5 kil. of ether at 3 atmospheres of pressure, or 440 litres of vapor. These cannot be in the ether condenser less than an atmosphere of resisting pressure; for the boiling point of ether is 38 deg.; so that expansion cannot be more than 3 atmospheres to 1.

The work of the ether cylinder will be

$$0.440 \times 3\ (1 + e \log. 3) - 1.32 \times 1 = 1.45.$$

If, instead of using the steam to vaporize ether, expansion is effected in a second cylinder, nearly up to the condenser pressure, it could be raised from 0.5 to 0.2 or 2.5 volumes; even supposing the most unfavorable conditions, as a tension of half an atmosphere at discharge from the first cylinder; which would imply a previous considerable expansion.

A kilogramme of steam at half an atmosphere, represents 3.200 mc.; hence the work would be

$$3.200 \times 0.5 (1 + e \log. 2.5) - 8 \times 0.2 = 1.47.$$

The work is theoretically the same as that with ether; but in fact it would be more considerable, and it is obtained more simply, without the help of a dangerous fluid, without complicated vaporizers and condensers, and other disadvantages. It is but a slight advantage that the ether cylinder would have a volume of 1.32, while that of the compound engine has a volume of 8.

A disadvantage in the use of ether is, that though the boiling point is low, its tension in the condenser is high, so that what is gained in one way is lost in another. Besides, its vapor has considerable density.

Combined vapor engines have given good

results when comparison was made with engines in which steam has been badly utilized. They gave splendid results when their consumption was 1.50 kil. per H. P., as compared with 2.25 to 2.50 kil. of steam engines. Nowadays, a kilogramme per H. P. is realized with much simpler means. Still it would not be fair to forget that ether engines have rendered important service by familiarizing us with the use of surface condensers. The majority of these engines, after the use of ether was given up, have been worked as steam engines with the use of ether vaporizers and condensers as surface condensers.

V. Our investigations naturally lead to the question whether by increase of pressures and expansions indefinite improvements may be realized; or, as Siemens has said in his report to the Institute of Mechanical Engineers, whether, in the course of ten years, we may not hope for a reduction of 50 per cent. in consumption.

We must first say a few words of the methods of determining the useful effect ot our engines. It is hardly necessary to remark upon the superficial and crude nature of the results generally given concerning the performance of a steam engine, results indicating the amount of fuel burned per hour, the unit of work upon pistons and cranks.

We often see the consumption indicated with the minutest accuracy in figures containing two or three decimals; and this without the slightest mention of the kind of fuel employed, though it is well known that the heating powers of combustibles vary greatly, so that an engine can run more economically with 2 kil. of one sort of fuel than with 1.50 kil. of another. What value have such reports?

Suppose the kind of fuel accurately stated; still the figure of consumption per H. P. has no scientific value because it confounds in one estimate a set of distinct elements which should be separately examined, as the amount of steam and its use, the work of the generator and that of the engine.

This is clear. Take for example an engine that uses 10 kil. of steam per H. P.

This is supplied by a boiler set under very bad conditions, vaporizing only 5 kilog. of water per kilog. of fuel. The consumption is 2 kil. hourly per H. P. Another engine may use 20 kilog. of steam, but if it is fed by a generator that very perfectly vaporizes 10 kilog. of water per kil. of fuel, the gross result of consumption per H. P. will also be 2 kilog. There is no apparent difference between the two.

The careful examination of this question is of great interest, for it leads to an important result: the conclusion that, if we give to the better of the two engines the better generator; or, if to the generator is attached an engine that utilizes the steam in the best manner, then, instead of 2 kilog., we should have 1 kilog. of consumption, or 50 per cent. economy.

That steam engines have been gradually brought to their present perfection, is due to the minute and careful analysis of the physical phenomena which convert the heat of coal into a motor force. It is easy to indicate the necessity of the separate study of the elements of the steam engine, but not easy to represent this separation in practical experiments. But it is possible, in land engines, to take a satisfactory account of the interior action of the apparatus, both by observation of the weight of fuel burned and of water converted into steam, and by measuring the work on the shaft by the Prony brake and the work upon the piston by the indicator. But in the case of great marine engines, the weight of fuel and the work upon the pistons are the only elements easily measured. Results must be obtained by comparison or by approximation, somewhat coarse, it is true, but yet of great interest, and with much probability of correctness in them, because of the great number of facts we have at hand.

In marine engines, the expenditure of steam—and therefore the performance—is generally estimated by what is termed the weight of sensible steam in the cylinders; that is, the weight as shown by the indicator diagrams. As we have seen, this indication is often illusory. Though in engines in which the condensation in the cylinders is limited by suitable conditions of action, it may approach the real figure, so as to give results accurate enough in practice, still, in other cases, it gives results of no possible value.

This weight of sensible steam in the cylinder, which we shall call P′, differs from the weight P′ of water actually vaporized in the boiler, because it does not take into account the loss of steam by leakages, condensations in the tubes or at the entrance to the cylinder, by dead spaces, etc. But it also differs from the weight p of steam theoretically necessary for a given pressure and expansion; because it takes into account the various restrictive elements of the motor force, such as the reduction of pressure between the boiler and the cylinder, the phases of distribution, lead, compression, actual expansion, obstruction, etc. Hence, in general

$$P > P' > p.$$

Further on we shall find apparent exceptions, for which we should be prepared.

We shall first determine, for various pressures and expansions, the theoretic weight of the vapor p consumed hourly to produce the work of a H. P., on the piston.

The following table gives the values of p for several kinds of marine engines. The figures are *theoretic*, being obtained from the formula for the work of steam. Only the resisting pressure in the condenser, estimated at 2 metres of water, is taken into account. (See page 64.)

The figure illustrates the results graphically. (See Fig. 5, page 65.)

The weight P′ of sensible steam in the cylinders shown by the indicator, should be more than the theoretic weight p, since the steam loses in various ways during its work. Of these losses, some are shown by the diagram, others are not sensible and introduce so many errors into the results that they are really not of practical use. They are almost exact enough for engines in which the condensation in the cylinders is a minimum. But they answer only when comparison is made of engines whose actions are nearly identical. Useful results may be obtained respecting the distribution of the

EXPANSION.	MODE OF ACTION.			
	Low pressure engines, 1½ atmosphere boiler pressure.	Mean pressure engines, 2½ atmospheres boiler pressure.	High pressure engines, 5 atmospheres boiler pressure.	Very high pressure engines, 10 atmospheres boiler pressure.
Volumes.	kg.	kg.	kg.	kg.
1.2	14 50	.....	.....	....
1.33	13.33	12.10	.....	....
1.50	12.20	11 05	.....	....
2.	.....	9.16	8 31	....
2.50	.....	8.10	7 34	....
3	.....	7.42	6.75	....
4	.....	6.50	5.90	....
5	.....	.....	5.40	....
6	.....	.....	5.05	....
8	.....	.....	4.60	....
10	.....	.....	4.25	3.92
12	.....	.....	4.05	3.70
15	.....	.....	3.80	3.50
20	.....	.....	....	3.25

FIG. 5.

10 Atmosph.
5 Atmosph.
Volumes
Expansion
1
1.5
2
4
6
8
10
12
15
20
25
5
10
15

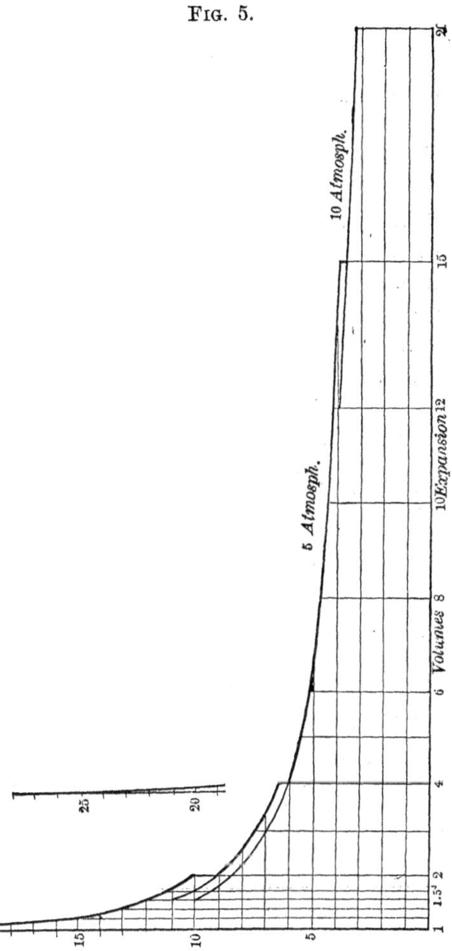

steam and the resistances in passages, pipes, etc.

The rate of the expansion normally employed ought to regulate the disposition of the parts of the engine; an engine constructed for a certain expansion will not work so well with a much greater expansion. The reasons have already been given; we add experimental confirmation as given in a table from "Engineering," in which is shown the weight of sensible steam in the cylinders of compound engines, each working with 3 deg. of expansion.

Number of engine.	1.	
Expansions	5 6	10
Absolute pressures	5.4 k.	5 k.
Velocity of piston	1.77 m.	1.49 m.
Weight of sens. steam in cyl per hour and per h. p.	6.38 k.	6.43 k.

Number of Engine.	2.	
Expansions	6.55	13.75
Absolute pressures	5.4 k.	5 k.
Velocity of Piston.	2.175 m.	1.75 m.
Weight of sens. steam in cyl. per hour and per h. p.	6.61 k.	6.43 k.

Numbea of Engine.	3.	
Expansions..........	7.4	16.74
Absolute pressures.	5.2 k.	4.7k.
Velocity of piston...........	1.98 m.	1.56 m.
Weight of sens. steam in cyl. per hour and per h. p.......	6.84 k.	7.29 k.

This table shows that the increase of expansion in the same engine increases instead of decreasing the consumption of steam. It is true that in the examples cited, action with full expansion corresponds to the least pressures and velocities of the pistons, and that there should result a diminution of useful effect of steam and an increase of losses by interior condensation. But the advantages due a greater expansion would more than compensate if it should act under normal conditions, precisely what it does not do. With an engine set to expand 5 or 6 volumes, expansions of 10, 12, or 15 volumes can be had only by a corresponding reduced admission into the first cylinder, so that one falls upon some of the inconveniences of single engines, to say nothing of the losses of pressure due to the intermediate spaces and passages of double cylinders.

We call attention to the ratio between the weight of sensible steam and that of the theoretic in order to warn the reader of an error which has escaped the notice of many persons. We annex a table. (See page 69.)

The first thing remarked in this table is that, while for most engines the sensible is greater than the theoretic weight, in other cases the ratio is one and less. Again, where this ratio is small, the consumption is increased, showing that the economy of steam is only apparent, there being no reason for supposing that the generation in the boiler is any less than in the other cases.

The conclusion is that we must not judge of the economy of an engine by the low rate of apparent expenditure of steam.

The explanation of this somewhat paradoxical result is simple, depending entirely upon what has been said above of the action of certain single engines without jackets. The vaporization during the period of expansion of the water condensed in the

Designation of engines.	Absolute boiler pressure.	Expansions.	Theoretic weight of steam per h. p.	Weight of sensible steam, P′, per h. p.	Ratio. $\frac{p'}{p}$.	Consumption of fuel per h. p.
	k.		k.	k.		k.
Woolf stationary engine.	5.15	10	4.25	5.60	1.30	0.95
Aigle, imperial yacht...	2.70	1.66	10	8.87	0.89	...
Abeille.............	2.12	1.54	11.30	10.50	0.90	2.25
Frigate, 250 h. p.......	2.47	1.54	11	11.10	1.00	1.75
Steamer, 600 h. p.......	2.10	2	9.50	10.79	1.14	1.78
Swift Steamer, 900 h. p.	1.98	1.66	10.70	12.00	1.12	1.64
Compound engines, 1...	5.40	5.60	5.20	6.38	1.22	....
Id. 2...	5.40	6.55	4.90	6.61	1.35	....
Id. 3...	5.20	7.40	4.70	6.84	1.57	....
Hercules, tug..........	2.17	1.66	10.35	13.00	1.25	....

cylinders, produces a total work greater than that which corresponds to the apparent weight of water in the cylinder; although this work is often much less than that which is due to the amount of steam from the boiler. Besides, the steam in the dead spaces expands during the period of expansion. The error is due to the fact that the diagram indicates only the presence in the cylinder of the vapor in a gaseous condition during the period of admission; that is, a part only of the steam which has entered the cylinder, which leaves it at the end of the stroke. One cannot be too careful in avoiding the errors which result from a superficial examination. We insist upon this point because it seems to have escaped the notice of writers, the majority of whom say nothing about it.

Though the value of the weight P′ is not of importance, it is otherwise with the quantity P, which represents the amount of water vaporized in the boiler. Unfortunately the direct measure of this quantity is attended with certain difficulties, and is possible only in a few cases. Especially in marine engines, where results are not to be obtained by experiment, we are obliged to deduce the weight of the fuel by assuming that we know the weight of water vaporized per kilogramme of fuel (an element which can be determined for a given fuel and generator), or to calculate it approximately in terms of p or P'.

In the first case, *i e.*, when P is deduced from the weight of fuel burned, it is necessary to take into account the water taken up by the steam, which tends to increase apparently the vaporization, and also to take into account the loss of heat by *extraction* when the boilers are fed with water containing salt. It is important to do this when comparison is made of engines condensing by injection with engines provided with surface condensers.

In land engines it is easier to measure directly the water introduced into the boiler, an element which should be determined as accurately as possible in experiments. In the example cited, of a Woolf beam engine, the weight of water vaporized per H. P. was found to be 6.5 kilo.; hence we have the following values:

P	6.50 k.	1.53	1.00
P′	5.60	1.30	0.85
p	4.25	1.00	0.653

At sea, where boilers of the same kind and action may be assumed to have almost identical vaporization, P may be estimated from the amount of coal burned. For example, we may assume a vaporization of 8 kil. of water per kil. of fuel, which is not too high an estimate for good coal, even with salt-water feed. The value of $\frac{P}{p}$ may then be deduced, and a sufficient number of data would determine the value of P in terms of p.

An examination of results from a very great number of engines, gives a mean ratio of 1.55—nearly the same amount as that obtained by direct measurement; so that we may safely assume the value of $\frac{P}{p}$ between 1.5 and 1.6. In some cases, it is true, this ratio is 1.3, and even 1.25; in others, it is 2. We must conclude that in the first case the vaporization is a little more than our assumed figure (8 kil.), and that, in the second case, there are special causes of the loss of steam. We do not give 1.5 or 1.6 as an exact coefficient; but we think it may be admitted as a mean value, and employed in *projets*.

A question occurs. The 50 or 60 per cent. which should be added to the theoretic weight of steam expended per hour to produce a H. P. of 75 kilogrammetres upon the piston, represents losses of all kinds, from boiler to condenser. Can these losses be reduced by suitable precaution and care in the construction of engines? The answer is, that it is to these reductions, and not to increase of expansion and tension, that we are to look for future improvements. It would be a great gain to reduce $\frac{P}{p}$ to 1.40 or 1.33.

We find that there is little to gain in the way of expansion; and possible increase of tension is very limited, on account of the difficulties in disposition of boilers, and those attendant upon too high a temperature. It does not seem possible, theoretically, to expend less than 3.50 kil. of steam; assume, rigorously, 3.25 kil., answering to a tension of 10 atm. and an expansion of 20 volumes. We suppose the most conditions perfectly favorable.

For the ratio $\frac{P}{p} = 1.50$, the weight of water to be converted into steam per H. P. is about 4.90 k. This, with a vaporization of 8.5 k. per kilogramme of fuel (with surface condensation), would lead to a consumption of 0.58 k. per H. P. per hour.

With the ratio $\frac{P}{p} = 1.33$, there would be the same result, with a theoretic weight of steam 3.70 kil., answering to an expansion of 12 volumes only to a pressure of 10 atmospheres; an expansion much easier to effect, and requiring dispositions much less complex than expansions of 20 volumes. Were it possible to combine the reduction of the ratio with the minimum expansion, the consumption would be reduced to 0.500 k.

So far, we have sought only for improvements in making use of the steam, without regard to the method of its generation. It is obvious that progress may be made in this direction, and that it may be considerable. For example, if the production of steam per kil. of fuel could be brought to from 8.5 k. to 10 k.—a result not hopeless—the expenditures would be reduced to 0.85; becoming 0.50 k., in the first case, and 0.43 in the second.

Such results will certainly not be obtained without great efforts; but we may reasonably hope to approximate them with means already at our disposal, while, in our opinion, it would be chimerical to attempt directly to reduce consumption by means of radical modifications in the generation and use of steam.

An objection is often urged, that it is indeed possible to make economic engines, but on condition of employing costly and cumbrous devices, too heavy and bulky.

A steam-engine consists of a generator to produce the motive vapor, and an engine to put it to use. The weight of an engine is in a certain measure controlled by its proper disposition, especially with regard to rapidity of action, while the weight of the generator necessarily depends only on the generation of steam and the pressure at which the steam is produced. The weight of the generator includes the weight of the boiler and the water it contains.

Boilers may be referred to various elements; as to the unit of power, the unit of grate surface, the unit of heating surface, etc.

For our purpose the unit of weight of fuel consumed is convenient; but care must be taken not to draw from the figures conclusions which they do not warrant. It is obvious that the weight of the boiler, per kilogramme of fuel, will be less in proportion to the amount of fuel burned every hour, and hence, that this weight will vary with the activity of combustion, and will not in any way give the measure of economical use of fuel and metal; hence the considerable differences in boilers of the same type. The true method of determining the efficiency of generators, with respect to weight, is to refer to the kilogramme of steam—if only one could get direct data of this element; but as this must generally be deduced from the weight of fuel, it is as well to make direct reference to it.

We prefer, in all cases, to refer to the weight of coal burned, rather than to the square metre of grate surface. This would be rigorously possible only in case the boilers burned the same quantity of coal; but as the consumption may vary within wide limits without a corresponding sensible change in the production of steam answering to a kilogramme of fuel, we may ask, what is the real value of such a mode of estimation?

With the above reservations, our method of measurement will permit us to take account of the possible results of a given kind of boiler.

The following table contains results corresponding to various kinds of boilers. (See page 78 for table.)

DESIGNATION.	Weight in kilogrammes of coal per hour.			KIND OF BOILER.
	Of empty boilers.	Water.	Of full boilers.	
	k.	k.	k.	
North, Crampton........	15.2	5.8	21	Locomotive.
North, 4 cylinders.......	24.6	7.4	32	Id.
North, moyennes Creusot	33.5	9.5	43	Id.
Francoise Ier......	28	20	48	Cylindric tubular. cast steel.
Machine compound......	58	17	75	Cylindric tubular.
Steamer, 900 h. p........	49	26	75	Regulation.
Steamer, 800 h. p.......	31	22	53	Id.
Ulster................	36	26.5	62.5	Tubular return flue.
Tasmanian..............	43	32	75	Id.
Sphynx................	68	41	109	With galleries.
Eldorado.........	55	50	105	Id.
Roanoke...............	76	42	118	Martin.
Brooklyn	65	51	116	Id.
Donawerth	35	24	59	Lamb and Summers.
Colombo...............	75	55	120	Id.
Guayaquil..............	61	51	112	Spiral, Randolph and Elder.
Hirondelle	31	2	33	Belleville.

Because of the tendency toward high pressures, it is probable that there will be a return to that type of locomotive boilers, which easily support pressures of 10 atmospheres, and which being fed with fresh water will give rise to no difficulty; always, with the condition of their having sufficient draft. It is clear that one of the first steps in improvement of marine boilers must be the adoption of means for producing an energetic draft. These boilers weigh at most, 43 kil. per kil. of fuel consumed; in some cases, 20 to 30 kil.; 40 kil. may be taken as an average. Another solution may consist in the use of the Belleville boilers, which, on the Hirondelle, gave 33 kil. But this figure should probably be increased on account of the fact that the coal was not well utilized. At any rate, we must admit the existence of boilers which do not weigh, water included, more than 40 kil. per kil. of fuel consumed hourly. A boiler of this kind feeding an engine expending only 0.5 k. per H. P., would weigh only 20 kil., water included, for the same unit.

Observing that the lighter boilers in combination with the most economic engines weigh 60 kil. per H. P., and the total, 80 to 100 kil., and more, we see that an economy of weight of 40 kil. at least, and possibly from 60 to 80 per H. P. is realized.

This weight, referred to an engine which, for a velocity of 60 revolutions per minute, weighs 80 to 100 kil. per H. P., represents a bonus of 50 per cent., which would allow an increase of the weight of the cylinders and their accessories; for it must not be forgotten that the excess of weight due to the increase of expansion affects only those parts on which the steam acts, and not the parts that transmit motion; while the parts that relate to condensation follow the reduction of the weight of the boilers. It is easy to verify this assertion rigorously. The development of the power in every steam-engine corresponds to a certain volume described by the pistons in a unit of time. This volume varies with the tension of the steam, the degree of expansion, and the efficiency of action of the engine.

The volume for a unit of power is found by multiplying the volume of the cylinders by twice the number of revolutions in a unit of time, and dividing this product by the indicated power. It can also be directly obtained without knowledge of the dimensions of the engine, by dividing the value of a unit of power by the mean effective pressure on the pistons, as shown by the indicator.

The volume, V, per H. P. is

$$\frac{\pi r^2 \, 2 c n}{P};$$

P being the power; but as $P = \pi r^2 \, 2 c n p$, p being the mean effective pressure of the steam, we have $V = \frac{1}{p}$.

It is more convenient to refer this to the minute. We have calculated for a great number of engines of all kinds.

In certain engines, in which builders have thought that they obtained lightness, they have sacrificed the performance; the volume per H. P. and per minute falling to 0.250 cm., and even to 0.200 cm. In most engines of ordinary action with tension at 2.5 k. and 2.75 k., this volume is 0.300 m. c. to 350 m. c.

In the compound engines, mentioned by

Bramwell, acting with full expansion, but at higher tensions, we find 0.385 cm. and 0.658 cm., as extreme values; the mean being 0.450. In engines with less expansion the values range from 0.300 to 0.400.

We infer that for the same number of revolutions, the volume of the cylinders will be increased only in the ratio of 1 to 1.5, an increase of volume corresponding to an increase in weight from 1.25 to 1.30. But suppose the weight rises to 1.50 or 2, so as to take into account the presence of jackets and the increase in section of slides and ports. In stamping engines, nowadays much used, the average weight of the cylinders with pistons and slides is 30 per cent. of the whole engine; it follows that, in this case, for a weight of 100 kil. per H. P., we should have 130 kil., or an increase of 30 kil.

We have seen that the minimum economy for boiler is 40 kil. We infer that a reduction of boiler-weight due to better use of steam, will permit the increase in weight of cylinders and pistons required by increase of expansion.

VALUABLE

SCIENTIFIC BOOKS,

PUBLISHED BY

D. VAN NOSTRAND,

23 MURRAY STREET AND 27 WARREN STREET,

NEW YORK.

FRANCIS. Lowell Hydraulic Experiments, being a selection from Experiments on Hydraulic Motors, on the Flow of Water over Weirs, in Open Canals of Uniform Rectangular Section, and through submerged Orifices and diverging Tubes. Made at Lowell, Massachusetts. By James B. Francis, C. E. 2d edition, revised and enlarged, with many new experiments, and illustrated with twenty-three copperplate engravings. 1 vol. 4to, cloth......................$15 00

ROEBLING (J. A.) Long and Short Span Railway Bridges. By John A. Roebling, C. E. Illustrated with large copperplate engravings of plans and views. Imperial folio, cloth............................ 25 00

CLARKE (T. C.) Description of the Iron Railway Bridge over the Mississippi River, at Quincy, Illinois. Thomas Curtis Clarke, Chief Engineer. Illustrated with 21 lithographed plans. 1 vol. 4to, cloth.. 7 50

TUNNER (P.) A Treatise on Roll-Turning for the Manufacture of Iron. By Peter Tunner. Translated and adapted by John B. Pearse, of the Pennsylvania Steel Works, with numerous engravings wood cuts and folio atlas of plates.................$10 00

ISHERWOOD (B. F.) Engineering Precedents for Steam Machinery. Arranged in the most practical and useful manner for Engineers. By B. F. Isherwood, Civil Engineer, U. S. Navy. With Illustrations. Two volumes in one. 8vo, cloth........... $2 50

BAUERMAN. Treatise on the Metallurgy of Iron, containing outlines of the History of Iron Manufacture, methods of Assay, and analysis of Iron Ores, processes of manufacture of Iron and Steel, etc., etc. By H. Bauerman. First American edition. Revised and enlarged, with an Appendix on the Martin Process for making Steel, from the report of Abram S. Hewitt. Illustrated with numerous wood engravings. 12mo, cloth.. 2 00

CAMPIN on the Construction of Iron Roofs. By Francis Campin. 8vo, with plates, cloth........... 3 00

COLLINS. The Private Book of Useful Alloys and Memoranda for Goldsmiths, Jewellers, &c. By James E. Collins. 18mo, cloth.................... 75

CIPHER AND SECRET LETTER AND TELEGRAPHIC CODE, with Hogg's Improvements. The most perfect secret code ever invented or discovered. Impossible to read without the key. By C. S. Larrabee. 18mo, cloth...................... 1 00

COLBURN. The Gas Works of London. By Zerah Colburn, C. E. 1 vol. 12mo, boards.............. 60

CRAIG (B. F.) Weights and Measures. An account of the Decimal System, with Tables of Conversion for Commercial and Scientific Uses. By B. F. Craig, M.D. 1 vol. square 32mo, limp cloth.... 50

NUGENT. Treatise on Optics; or, Light and Sight, theoretically and practically treated; with the application to Fine Art and Industrial Pursuits. By E. Nugent. With one hundred and three illustrations. 12mo, cloth.................................... 2 00

GLYNN (J.) Treatise on the Power of Water, as applied to drive Flour Mills, and to give motion to Turbines and other Hydrostatic Engines. By Joseph

Glynn. Third edition, revised and enlarged, with numerous illustrations. 12mo, cloth.............. $1 00

HUMBER. A Handy Book for the Calculation of Strains in Girders and similar Structures, and their Strength, consisting of Formulæ and corresponding Diagrams, with numerous details for practical application. By William Humber. 12mo, fully illustrated, cloth.................................. 2 50

GRUNER. The Manufacture of Steel. By M. L. Gruner. Translated from the French, by Lenox Smith, with an appendix on the Bessamer process in the United States, by the translator. Illustrated by Lithographed drawings and wood cuts. 8vo, cloth.. 3 50

AUCHINCLOSS. Link and Valve Motions Simplified. Illustrated with 37 wood-cuts, and 21 lithographic plates, together with a Travel Scale, and numerous useful Tables. By W. S. Auchincloss. 8vo, cloth.. 3 00

VAN BUREN. Investigations of Formulas, for the strength of the Iron parts of Steam Machinery. By J. D. Van Buren, Jr., C. E. Illustrated, 8vo, cloth. 2 00

JOYNSON. Designing and Construction of Machine Gearing. Illustrated, 8vo, cloth.................. 2 00

GILLMORE. Coignet Beton and other Artificial Stone. By Q. A. Gillmore, Major U. S. Corps Engineers. 9 plates, views, &c. 8vo, cloth.................... 2 50

SAELTZER. Treattse on Acoustics in connection with Ventilation. By Alexander Saeltzer, Architect. 12mo, cloth...................................... 2 00

THE EARTH'S CRUST. A handy Outline of Geology. By David Page. Illustrated, 18mo, cloth.... 75

DICTIONARY of Manufactures, Mining, Machinery, and the Industrial Arts. By George Dodd. 12mo, cloth.. 2 00

FRANCIS. On the Strength of Cast-Iron Pillars, with Tables for the use of Engineers, Architects, and Builders. By James B. Francis, Civil Engineer. 1 vol. 8vo, cloth................................ 2 00

GILLMORE (Gen. Q. A.) Treatise on Limes, Hydraulic Cements, and Mortars. Papers on Practical Engineering, U. S. Engineer Department, No. 9, containing Reports of numerous Experiments conducted in New York City, during the years 1858 to 1861, inclusive. By Q. A. Gillmore, Bvt. Maj.-Gen., U. S. A., Major, Corps of Engineers. With numerous illustrations. 1 vol, 8vo, cloth............... $4 00

HARRISON. The Mechanic's Tool Book, with Practical Rules and Suggestions for Use of Machinists, Iron Workers, and others. By W. B. Harrison, associate editor of the "American Artisan." Illustrated with 44 engravings. 12mo, cloth............ 1 50

HENRICI (Olaus). Skeleton Structures, especially in their application to the Building of Steel and Iron Bridges. By Olaus Henrici. With folding plates and diagrams. 1 vol. 8vo, cloth................... 3 00

HEWSON (Wm.) Principles and Practice of Embanking Lands from River Floods, as applied to the Levees of the Mississippi. By William Hewson, Civil Engineer. 1 vol. 8vo, cloth....................... 2 00

HOLLEY (A. L.) Railway Practice. American and European Railway Practice, in the economical Generation of Steam, including the Materials and Construction of Coal-burning Boilers, Combustion, the Variable Blast, Vaporization, Circulation, Superheating, Supplying and Heating Feed-water, etc., and the Adaptation of Wood and Coke-burning Engines to Coal-burning; and in Permanent Way, including Road-bed, Sleepers, Rails, Joint-fastenings, Street Railways, etc., etc. By Alexander L. Holley, B. P. With 77 lithographed plates. 1 vol. folio, cloth.... 12 00

KING (W. H.) Lessons and Practical Notes on Steam, the Steam Engine, Propellers, etc., etc., for Young Marine Engineers, Students, and others. By the late W. H. King, U. S. Navy. Revised by Chief Engineer J. W. King, U. S. Navy. Twelfth edition, enlarged. 8vo, cloth. 2 00

MINIFIE (Wm.) Mechanical Drawing. A Text-Book of Geometrical Drawing for the use of Mechanics and Schools, in which the Definitions and Rules of Geometry are familiarly explained; the Practical Problems are arranged, from the most simple to the more complex, and in their description technicalities are avoided as much as possible. With illustrations for Drawing Plans, Sections, and Elevations of Railways and Machinery; an Introduction to Isometrical Drawing, and an Essay on Linear Perspective and Shadows. Illustrated with over 200 diagrams engraved on steel. By Wm. Minifie, Architect. Seventh edition. With an Appendix on the Theory and Application of Colors. 1 vol. 8vo, cloth.......... $4 00

"It is the best work on Drawing that we have ever seen, and is especially a text-book of Geometrical Drawing for the use of Mechanics and Schools. No young Mechanic, such as a Machinists, Engineer, Cabinet-maker, Millwright, or Carpenter, should be without it."—*Scientific American.*

—— Geometrical Drawing. Abridged from the octavo edition, for the use of Schools. Illustrated with 48 steel plates. Fifth edition. 1 vol. 12mo, cloth.... 2 00

STILLMAN (Paul.) Steam Engine Indicator, and the Improved Manometer Steam and Vacuum Gauges—their Utility and Application. By Paul Stillman. New edition. 1 vol. 12mo, flexible cloth........... 1 00

SWEET (S. H.) Special Report on Coal; showing its Distribution, Classification, and cost delivered over different routes to various points in the State of New York, and the principal cities on the Atlantic Coast. By S. H. Sweet. With maps, 1 vol. 8vo, cloth..... 3 00

WALKER (W. H.) Screw Propulsion. Notes on Screw Propulsion: its Rise and History. By Capt. W. H. Walker, U. S. Navy. 1 vol. 8vo, cloth..... 75

WARD (J. H.) Steam for the Million. A popular Treatise on Steam and its Application to the Useful Arts, especially to Navigation. By J. H. Ward, Commander U. S. Navy. New and revised edition. 1 vol. 8vo, cloth................................ 1 00

WEISBACH (Julius). Principles of the Mechanics of Machinery and Engineering. By Dr. Julius Weisbach, of Freiburg. Translated from the last German edition. Vol. I., 8vo, cloth.. .... 10 00

DIEDRICH. The Theory of Strains, a Compendium for the calculation and construction of Bridges, Roofs, and Cranes, with the application of Trigonometrical Notes, containing the most comprehensive information in regard to the Resulting strains for a permanent Load, as also for a combined (Permanent and Rolling) Load. In two sections, adadted to the requirements of the present time. By John Diedrich, C. E. Illustrated by numerous plates and diagrams. 8vo, cloth.................................... 5 00

WILLIAMSON (R. S.) On the use of the Barometer on Surveys and Reconnoissances. Part I. Meteorology in its Connection with Hypsometry. Part II. Barometric Hypsometry. By R. S. Wiliamson, Bvt. Lieut.-Col. U. S. A., Major Corps of Engineers. With Illustrative Tables and Engravings. Paper No. 15, Professional Papers, Corps of Engineers. 1 vol. 4to, cloth.................................... 15 00

POOK (S. M.) Method of Comparing the Lines and Draughting Vessels Propelled by Sail or Steam. Including a chapter on Laying off on the Mould-Loft Floor. By Samuel M. Pook, Naval Constructor. 1 vol. 8vo, with illustrations, cloth........... 5 00

ALEXANDER (J. H.) Universal Dictionary of Weights and Measures, Ancient and Modern, reduced to the standards of the United States of America. By J. H. Alexander. New edition, enlarged. 1 vol. 8vo, cloth.................................... 3 50

BROOKLYN WATER WORKS. Containing a Descriptive Account of the Construction of the Works, and also Reports on the Brooklyn, Hartford, Belleville and Cambridge Pumping Engines. With illustrations. 1 vol. folio, cloth...........................

RICHARDS' INDICATOR. A Treatise on the Richards Steam Engine Indicator, with an Appendix by F. W. Bacon, M. E. 18mo, flexible, cloth.......... 1 00

POPE. Modern Practice of the Electric Telegraph. A Hand Book for Electricians and operators. By Frank L. Pope. Eighth edition, revised and enlarged, and fully illustrated. 8vo, cloth........................ $2.00

"There is no other work of this kind in the English language that contains in so small a compass so much practical information in the application of galvanic electricity to telegraphy. It should be in the hands of every one interested in telegraphy, or the use of Batteries for other purposes."

MORSE. Examination of the Telegraphic Apparatus and the Processes in Telegraphy. By Samuel F. Morse, LL.D., U. S. Commissioner Paris Universal Exposition, 1867. Illustrated, 8vo, cloth........... $2 00

SABINE. History and Progress of the Electric Telegraph, with descriptions of some of the apparatus. By Robert Sabine, C. E. Second edition, with additions, 12mo, cloth.................................. 1 25

CULLEY. A Hand-Book of Practical Telegraphy. By R. S. Culley, Engineer to the Electric and International Telegraph Company. Fourth edition, revised and enlarged. Illustrated 8vo, cloth.............. 5 00

BENET. Electro-Ballistic Machines, and the Schultz Chronoscope. By Lieut.-Col. S. V. Benet, Captain of Ordnance, U. S. Army. Illustrated, second edition, 4to, cloth.................................. 3 00

MICHAELIS. The Le Boulenge Chronograph, with three Lithograph folding plates of illustrations. By Brevet Captain O. E. Michaelis, First Lieutenant Ordnance Corps, U. S. Army, 4to, cloth.......... 3 00

ENGINEERING FACTS AND FIGURES. An Annual Register of Progress in Mechanical Engineering and Construction. for the years 1863, 64, 65, 66, 67, 68. Fully illustrated, 6 vols. 18mo, cloth, $2.50 per vol., each volume sold separately..............

HAMILTON. Useful Information for Railway Men. Compiled by W. G. Hamilton, Engineer. Fifth edition, revised and enlarged, 562 pages Pocket form. Morocco, gilt.................................. 2 00

STUART. The Civil and Military Engineers of America. By Gen. C. B. Stuart. With 9 finely executed portraits of eminent engineers, and illustrated by engravings of some of the most important works constructed in America. 8vo, cloth.................. $5 00

STONEY. The Theory of Strains in Girders and similar structures, with observations on the application of Theory to Practice, and Tables of Strength and other properties of Materials. By Bindon B. Stoney, B. A. New and revised edition, enlarged, with numerous engravings on wood, by Oldham. Royal 8vo, 664 pages. Complete in one volume. 8vo, cloth...... 15 00

SHREVE. A Treatise on the Strength of Bridges and Roofs. Comprising the determination of Algebraic formulas for strains in Horizontal, Inclined or Rafter, Triangular, Bowstring, Lenticular and other Trusses, from fixed and moving loads, with practical applications and examples, for the use of Students and Engineers. By Samuel H. Shreve, A. M., Civil Engineer. 87 wood cut illustrations. 8vo, cloth.............. 5 00

MERRILL. Iron Truss Bridges for Railroads. The method of calculating strains in Trusses, with a careful comparison of the most prominent Trusses, in reference to economy in combination, etc., etc. By Brevet. Col. William E. Merrill, U S. A., Major Corps of Engineers, with nine lithographed plates of Illustrations. 4to, cloth.......................... 5 00

WHIPPLE. An Elementary and Practical Treatise on Bridge Building. An enlarged and improved edition of the author's original work. By S. Whipple, C. E., inventor of the Whipple Bridges, &c. Illustrated 8vo, cloth...................................... 4 00

THE KANSAS CITY BRIDGE. With an account of the Regimen of the Missouri River, and a description of the methods used for Founding in that River. By O. Chanute, Chief Engineer, and George Morrison, Assistant Engineer. Illustrated with five lithographic views and twelve plates of plans. 4to, cloth, 6 00

MAC CORD. A Practical Treatise on the Slide Valve by Eccentrics, examining by methods the action of the Eccentric upon the Slide Valve, and explaining the Practical processes of laying out the movements, adapting the valve for its various duties in the steam engine. For the use of Engineers, Draughtsmen, Machinists, and Students of Valve Motions in general. By C. W. Mac Cord, A. M., Professor of Mechanical Drawing, Stevens' Institute of Technology, Hoboken, N. J. Illustrated by 8 full page copper-plates. 4to, cloth................................ $4 00

KIRKWOOD. Report on the Filtration of River Waters, for the supply of cities, as practised in Europe, made to the Board of Water Commissioners of the City of St. Louis. By James P. Kirkwood. Illustrated by 30 double plate engravings. 4to, cloth, 15 00

PLATTNER. Manual of Qualitative and Quantitative Analysis with the Blow Pipe. From the last German edition, revised and enlarged. By Prof. Th. Richter, of the Royal Saxon Mining Academy. Translated by Prof. H. B. Cornwall, Assistant in the Columbia School of Mines, New York assisted by John H. Caswell. Illustrated with 87 wood cuts, and one lithographic plate. Second edition, revised, 560 pages, 8vo, cloth.................................. 7 50

PLYMPTON. The Blow Pipe. A system of Instruction in its practical use being a graduated course of analysis for the use of students, and all those engaged in the examination of metallic combinations Second edition, with an appendix and a copious index. By Prof. Geo W. Plympton, of the Polytechnic Institute, Brooklyn, N. Y. 12mo, cloth................ 2 00

PYNCHON. Introduction to Chemical Physics, designed for the use of Academies, Colleges and High Schools. Illustrated with numerous engravings, and containing copious experiments with directions for preparing them. By Thomas Ruggles Pynchon, M A., Professor of Chemistry and the Natural Sciences, Trinity College, Hartford New edition, revised and enlarged and illustrated by 269 illustrations on wood. Crown, 8vo. cloth.......................... 3 00

ELIOT AND STORER. A compendious Manual of Qualitative Chemical Analysis. By Charles W. Eliot and Frank H. Storer. Revised with the Co-operation of the authors. By William R. Nichols, Professor of Chemistry in the Massachusetts Institute of Technology Illustrated, 12mo, cloth....... $1 50

RAMMELSBERG. Guide to a course of Quantitative Chemical Analysis, especially of Minerals and Furnace Products. Illustrated by Examples By C. F. Rammelsberg. Translated by J. Towler, M. D. 8vo, cloth.. 2 25

EGLESTON. Lectures on Descriptive Mineralogy, delivered at the School of Mines, Columbia College. By Professor T. Egleston. Illustrated by 34 Lithographic Plates. 8vo, cloth........................ 4 50

MITCHELL. A Manual of Practical Assaying. By John Mitchell. Third edition. Edited by William Crookes, F. R. S. 8vo, cloth........................ 10 00

WATT'S Dictionary of Chemistry. New and Revised edition complete in 6 vols 8vo cloth, $62.00. Supplementary volume sold separately. Price, cloth... 9 00

RANDALL. Quartz Operators Hand-Book. By P. M. Randall. New edition, revised and enlarged, fully illustrated. 12mo, cloth........................ 2 00

SILVERSMITH. A Practical Hand-Book for Miners, Metallurgists, and Assayers, comprising the most recent improvements in the disintegration amalgamation, smelting, and parting of the Precious ores, with a comprehensive Digest of the Mining Laws. Greatly augmented, revised and corrected. By Julius Silversmith. Fourth edition. Profusely illustrated. 12mo, cloth.. 3 00

THE USEFUL METALS AND THEIR ALLOYS, including Mining Ventilation, Mining Jurisprudence, and Metallurgic Chemistry employed in the conversion of Iron, Copper, Tin, Zinc, Antimony and Lead ores, with their applications to the Industrial Arts. By Scoffren, Truan, Clay, Oxland, Fairbairn, and others. Fifth edition, half calf..................... 3 75

JOYNSON. The Metals used in construction, Iron, Steel, Bessemer Metal, etc., etc. By F. H. Joynson. Illustrated, 12mo, cloth.......................... $0 75

VON COTTA. Treatise on Ore Deposits. By Bernhard Von Cotta, Professor of Geology in the Royal School of Mines, Freidberg, Saxony. Translated from the second German edition, by Frederick Prime, Jr., Mining Engineer, and revised by the author, with numerous illustrations. 8vo, cloth....... 4 00

URE. Dictionary of Arts, Manufactures and Mines. By Andrew Ure, M.D. Sixth edition, edited by Robert Hunt, F. R. S., greatly enlarged and re-written. London, 1872. 3 vols. 8vo, cloth, $25.00. Half Russia.................................... 37 50

BELL. Chemical Phenomena of Iron Smelting. An experimental and practical examination of the circumstances which determine the capacity of the Blast Furnace, The Temperature of the air, and the proper condition of the Materials to be operated upon. By I. Lowthian Bell. 8vo, cloth........... 6 00

ROGERS. The Geology of Pennsylvania. A Government survey, with a general view of the Geology of the United States, Essays on the Coal Formation and its Fossils, and a description of the Coal Fields of North America and Great Britain. By Henry Darwin Rogers, late State Geologist of Pennsylvania, Splendidly illustrated with Plates and Engravings in the text. 3 vols., 4to, cloth with Portfolio of Maps. 30 00

BURGH. Modern Marine Engineering, applied to Paddle and Screw Propulsion. Consisting of 36 colored plates, 259 Practical Wood Cut Illustrations, and 403 pages of descriptive matter, the whole being an exposition of the present practice of James Watt & Co., J. & G. Rennie, R. Napier & Sons, and other celebrated firms, by N. P. Burgh, Engineer, thick 4to, vol., cloth, $25.00; half mor........ 30 00

BARTOL. Treatise on the Marine Boilers of the United States. By B. H. Bartol. Illustrated, 8vo, cloth... 1 50

BOURNE. Treatise on the Steam Engine in its various applications to Mines, Mills, Steam Navigation, Railways, and Agriculture, with the theoretical investigations respecting the Motive Power of Heat, and the proper proportions of steam engines. Elaborate tables of the right dimensions of every part, and Practical Instructions for the manufacture and management of every species of Engine in actual use. By John Bourne, being the ninth edition of "A Treatise on the Steam Engine," by the "Artizan Club." Illustrated by 38 plates and 546 wood cuts. 4to, cloth.......................................$15 00

STUART. The Naval Dry Docks of the United States. By Charles B. Stuart late Engineer-in-Chief of the U. S. Navy. Illustrated with 24 engravings on steel. Fourth edition, cloth.................... 6 00

EADS. System of Naval Defences. By James B. Eads, C. E., with 10 illustrations, 4to, cloth........ 5 00

FOSTER. Submarine Blasting in Boston Harbor, Massachusetts. Removal of Tower and Corwin Rocks. By J. G. Foster, Lieut.-Col. of Engineers, U. S. Army. Illustrated with seven plates, 4to, cloth.. 3 50

BARNES Submarine Warfare, offensive and defensive, including a discussion of the offensive Torpedo System, its effects upon Iron Clad Ship Systems and influence upon future naval wars. By Lieut.-Commander J. S. Barnes, U. S. N., with twenty lithographic plates and many wood cuts. 8vo, cloth..... 5 00

HOLLEY. A Treatise on Ordnance and Armor, embracing descriptions, discussions, and professional opinions concerning the materials, fabrication, requirements, capabilities, and endurance of European and American Guns, for Naval, Sea Coast, and Iron Clad Warfare, and their Rifling, Projectiles, and Breech-Loading; also, results of experiments against armor, from official records, with an appendix referring to Gun Cotton, Hooped Guns, etc., etc. By Alexander L. Holley, B. P., 948 pages, 493 engravings, and 147 Tables of Results, etc., 8vo, half roan. 10 00

SIMMS. A Treatise on the Principles and Practice of Levelling, showing its application to purposes of Railway Engineering and the Construction of Roads, &c. By Frederick W. Simms, C. E. From the 5th London edition, revised and corrected, with the addition of Mr. Laws's Practical Examples for setting out Railway Curves. Illustrated with three Lithographic plates and numerous wood cuts. 8vo, cloth. $2 50

BURT. Key to the Solar Compass, and Surveyor's Companion; comprising all the rules necessary for use in the field; also description of the Linear Surveys and Public Land System of the United States, Notes on the Barometer, suggestions for an outfit for a survey of four months, etc. By W. A. Burt, U. S. Deputy Surveyor. Second edition. Pocket book form, tuck.. 2 50

THE PLANE TABLE. Its uses in Topographical Surveying, from the Papers of the U. S. Coast Survey. Illustrated, 8vo, cloth........................ 2 00

"This work gives a description of the Plane Table, employed at the U. S. Coast Survey office, and the manner of using it."

JEFFER'S. Nautical Surveying. By W. N. Jeffers, Captain U. S. Navy. Illustrated with 9 copperplates and 31 wood cut illustrations. 8vo, cloth........... 5 00

CHAUVENET. New method of correcting Lunar Distances, and improved method of Finding the error and rate of a chronometer, by equal altitudes. By W. Chauvenet, LL.D. 8vo, cloth................. 2 00

BRUNNOW. Spherical Astronomy. By F. Brunnow, Ph. Dr. Translated by the author from the second German edition. 8vo, cloth...................... 6 50

PEIRCE. System of Analytic Mechanics. By Benjamin Peirce. 4to, cloth.......................... 10 00

COFFIN. Navigation and Nautical Astronomy. Prepared for the use of the U. S. Naval Academy. By Prof. J. H. C. Coffin. Fifth edition. 52 wood cut illustrations. 12mo, cloth............................ 3 50

CLARK. Theoretical Navigation and Nautical Astronomy. By Lieut. Lewis Clark, U. S. N. Illustrated with 41 wood cuts. 8vo, cloth.................... $3 00

HASKINS. The Galvanometer and its Uses. A Manual for Electricians and Students. By C. H. Haskins. 12mo, pocket form, morocco. (In press).....

GOUGE. New System of Ventilation, which has been thoroughly tested, under the patronage of many distinguished persons. By Henry A. Gouge. With many illustrations. 8vo, cloth..................... 2 00

BECKWITH. Observations on the Materials and Manufacture of Terra-Cotta, Stone Ware, Fire Brick, Porcelain and Encaustic Tiles, with remarks on the products exhibited at the London International Exhibition, 1871. By Arthur Beckwith, C. E. 8vo, paper.. 60

MORFIT. A Practical Treatise on Pure Fertilizers, and the chemical conversion of Rock Guano, Marlstones, Coprolites, and the Crude Phosphates of Lime and Alumina generally, into various valuable products. By Campbell Morfit, M.D., with 28 illustrative plates, 8vo, cloth.................................... 20 00

BARNARD. The Metric System of Weights and Measures. An address delivered before the convocation of the University of the State of New York, at Albany, August, 1871. By F. A. P. Barnard, LL.D., President of Columbia College, New York. Second edition from the revised edition, printed for the Trustees of Columbia College. Tinted paper, 8vo, cloth 3 00

—— Report on Machinery and Processes on the Industrial Arts and Apparatus of the Exact Sciences. By F. A. P. Barnard, LL.D. Paris Universal Exposition, 1867. Illustrated, 8vo, cloth.............. 5 00

BARLOW. Tables of Squares, Cubes, Square Roots, Cube Roots, Reciprocals of all integer numbers up to 10,000. New edition, 12mo, cloth................. 2 50

MYER. Manual of Signals, for the use of Signal officers in the Field, and for Military and Naval Students, Military Schools, etc. A new edition enlarged and illustrated By Brig. General Albert J. Myer, Chief Signal Officer of the army, Colonel of the Signal Corps during the War of the Rebellion. 12mo, 48 plates, full Roan.................................. $5 00

WILLIAMSON. Practical Tables in Meteorology and Hypsometry, in connection with the use of the Barometer. By Col. R. S. Williamson, U. S. A. 4to, cloth.. 2 50

THE YOUNG MECHANIC. Containing directions for the use of all kinds of tools, and for the construction of Steam Engines and Mechanical Models, including the Art of Turning in Wood and Metal. By the author "The Lathe and its Uses," etc. From the English edition with corrections. Illustrated, 12mo, cloth.. 1 75

PICKERT AND METCALF. The Art of Graining. How Acquired and How Produced, with description of colors, and their application. By Charles Pickert and Abraham Metcalf. Beautifully illustrated with 42 tinted plates of the various woods used in interior finishing. Tinted paper, 4to, cloth................ 10 00

HUNT. Designs for the Gateways of the Southern Entrances to the Central Park. By Richard M. Hunt. With a description of the designs. 4to. cloth...... 5 00

LAZELLE. One Law in Nature. By Capt. H. M. Lazelle, U. S. A. A new Corpuscular Theory, comprehending Unity of Force, Identity of Matter, and its Multiple Atom Constitution, applied to the Physical Affections or Modes of Energy. 12mo, cloth... 1 50

PETERS. Notes on the Origin, Nature, Prevention, and Treatment of Asiatic Cholera. By John C. Peters, M. D. Second edition, with an Appendix. 12mo, cloth.. 1 50

BOYNTON. History of West Point, its Military Importance during the American Revolution, and the Origin and History of the U. S. Military Academy. By Bvt. Major C. E. Boynton, A.M., Adjutant of the Military Academy. Second edition, 416 pp. 8vo, printed on tinted paper, beautifully illustrated with 36 maps and fine engravings, chiefly from photographs taken on the spot by the author. Extra cloth.. $3 50

WOOD. West Point Scrap Book, being a collection of Legends, Stories, Songs, etc., of the U. S. Military Academy. By Lieut. O. E. Wood, U. S. A. Illustrated by 69 engravings and a copperplate map. Beautifully printed on tinted paper. 8vo, cloth..... 5 00

WEST POINT LIFE. A Poem read before the Dialectic Society of the United States Military Academy. Illustrated with Pen-and-Ink Sketches. By a Cadet. To which is added the song, "Benny Havens, oh!" oblong 8vo, 21 full page illustrations, cloth......... 2 50

GUIDE TO WEST POINT and the U. S. Military Academy, with maps and engravings, 18mo, blue cloth, flexible.................................. 1 00

HENRY. Military Record of Civilian Appointments in the United States Army By Guy V. Henry, Brevet Colonel and Captain First United States Artillery, Late Colonel and Brevet Brigadier General, United States Volunteers. Vol. 1 now ready. Vol. 2 in press. 8vo, per volume, cloth.................... 5 00

HAMERSLY. Records of Living Officers of the U. S. Navy and Marine Corps. Compiled from official sources. By Lewis R. Hamersly, late Lieutenant U. S. Marine Corps. Revised edition, 8vo, cloth... 5 00

MOORE. Portrait Gallery of the War. Civil, Military and Naval. A Biographical record, edited by Frank Moore. 60 fine portraits on steel. Royal 8vo, cloth.. 6 00

www.ingramcontent.com/pod-product-compliance
Lightning Source LLC
LaVergne TN
LVHW021430110826
845150LV00007B/2168

* 9 7 8 1 4 2 5 5 0 6 5 7 5 *